YOUR KNOWLEDGE HAS VALUE

Ankit Ponkia

Analysis and design of rectangular microstrip patch antenna on different substrate materials in X-Band

Antenna and wave propagation

GRIN Publishing

Bibliographic information published by the German National Library:

The German National Library lists this publication in the National Bibliography; detailed bibliographic data are available on the Internet at http://dnb.dnb.de .

Imprint:

Copyright © 2014 GRIN Verlag GmbH
Print and binding: Books on Demand GmbH, Norderstedt Germany
ISBN: 978-3-656-61680-1

This book at GRIN:

http://www.grin.com/en/e-book/270364/analysis-and-design-of-rectangular-microstrip-patch-antenna-on-different

GRIN - Your knowledge has value

Since its foundation in 1998, GRIN has specialized in publishing academic texts by students, college teachers and other academics as e-book and printed book. The website www.grin.com is an ideal platform for presenting term papers, final papers, scientific essays, dissertations and specialist books.

Visit us on the internet:

http://www.grin.com/

http://www.facebook.com/grincom

http://www.twitter.com/grin_com

Analysis and Design of Rectangular Microstrip Patch antenna on Different Substrate Materials in X-Band

Ankit V. Ponkia

Asst. Prof. Dept. of Electronics & Communication Engineering, Shantilal Shah Government Engineering College, Sidsar Campus, Bhavnagar-340460, Gujarat, India

Abstract— In this paper software based design and analysis has been carried out for a rectangular patch antenna using different substrate materials. A coaxial probe fed rectangular microstrip patch antenna operating at X-band (8 to 12 GHz) is analyzed on different substrate materials like Rogers RT/duroid 5880, Rogers RT/duroid 5870, Neltec NX9240, Arlon DiClad 522, and FR4_epoxy. The design is analyzed by Finite Element Method (FEM) based HFSS™ EM simulator software. Return loss, VSWR plot, smith chart and radiation pattern plots are observed and plotted for all antennas.

Index Terms—Rectangular Patch, Substrate Materials, Microstrip Antennas, Return Loss, VSWR.

INTRODUCTION

Microstrip patch antennas (MSAs) have attractive features of low profile, light weight and easy fabrication process. The reductions in size and bandwidth enhancement are some major design considerations for practical applications of microstrip antennas [1]. The developing markets like personal communication systems (PCS), mobile satellite communications and intelligent vehicle highway systems (IVHS) and many other suggest that the demand for microstrip antennas and arrays will increase even further [2]. Microstrip patch antenna (MSA) consist of radiating or conducting patch printed on one side of dielectric substrate material with ground plane on other side [3].There are so many types of radiating patch configuration are available such as the rectangular or square patch, triangular, circular disc, ellipse ,annular ring and pentagon [4]. Rectangular microstrip patch is much attracted due to their simple structure as it one of the simplest patch configuration. There are numerous substrates are available that can be used for the design of microstrip antennas, and their dielectric constants are usually in the range of 2.1 < εr < 25 [5-6]. A radiating patch is very thin (t << λ_0, where λ_0 is the free-space wavelength) metallic strip. The patch is generally made of conducting material e.g. copper. The substrate has thickness h (h << λ_0, usually $0.003\lambda_0$ << h << $0.05\lambda_0$, where λ_0 is free space wavelength) [7].

In this paper rectangular microstrip patch antena (RMSA) is analyzed on different dielectric substrate (or laminates) materials operating in X-band.

ANTENNA DESIGN

The rectangular patch is the most widely used patch configuration. It can be analyze using both the transmission-line and cavity models, which are most accurate for thin substrates [8]. The figure 1 shows geometry of proposed antenna.

For rectangular patch antenna practical approximate relation for the normalized extension of the length is given by [9]

$$\frac{\Delta L}{h} = 0.412 \frac{(\varepsilon_{reff} + 0.3)\left(\frac{W}{h} + 0.264\right)}{(\varepsilon_{reff} - 0.258)\left(\frac{W}{h} + 0.8\right)} \quad (1)$$

Where width-to-height ratio,

$$\frac{W}{h} > 1 \qquad (2)$$

Here effective dielectric constant ε_{reff} is given by [10],

$$\varepsilon_{reff} = \frac{\varepsilon_r + 1}{2} + \frac{\varepsilon_r - 1}{2}\left[1 + \frac{12h}{W}\right]^{-1/2} \qquad (3)$$

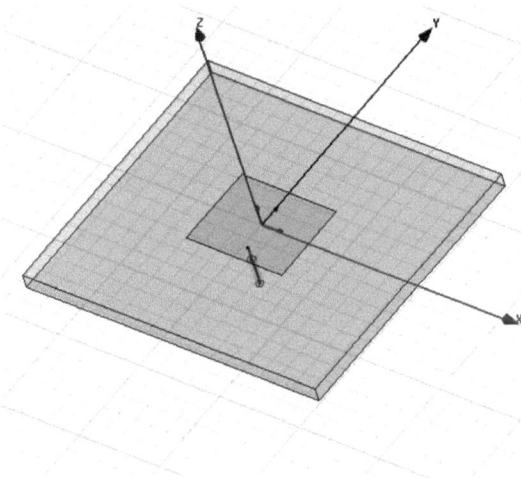

Figure 1: Geometry of antenna

Effective length of the patch is now given by,

$$L_{eff} = L + 2\Delta L \qquad (4)$$

For the dominant TM_{010} mode, the resonant frequency of the rectangular microstrip antenna is function of its actual length L which is given by,

$$(fr)_{010} = \frac{\vartheta_o}{2L\sqrt{\varepsilon_r}} \qquad (5)$$

Where ϑ_o speed of light. Equation (5) does not account for fringing so to include edge effects and considering fringing equation (5) can be written as,

$$(fr)_{010} = \frac{1}{2L_{eff}\sqrt{\varepsilon_{eff}\mu_0\varepsilon_0}} = \frac{1}{2(L + 2\Delta L)\sqrt{\varepsilon_{eff}\mu_0\varepsilon_0}} \qquad (6)$$

Design Procedure Steps:

1. Determine practical width of patch radiator, which is given by [3]

$$W = \frac{1}{2f_r}\frac{1}{\sqrt{\mu_0\varepsilon_0}}\sqrt{\frac{2}{\varepsilon_r+1}} = \frac{\vartheta_0}{2f_r}\sqrt{\frac{2}{\varepsilon_r+1}} \qquad (7)$$

2. Determine effective dielectric constant ε_{reff} from equation (3)
3. Determine the extension of the patch length ΔL using equation (1)
4. Finally, calculate actual length L by solving equation (6). L is given by,

$$L = \frac{1}{2f_r\sqrt{\varepsilon_{eff}\mu_0\varepsilon_0}} - 2\Delta L \qquad (8)$$

The geometry of rectangular microstrip antenna is shown in figure 1.The rectangular patch is mounted on different substrates materials like Rogers RT/duroid 5880, Rogers RT/duroid 5870, Neltec NX9240, Arlon DiClad 522, and FR4_epoxy. The table 1 shows substrate materials with dielectric constant .Excitation to patch conductor was given using wave port. The coaxial probe feeding technique is used. The substrate with dimension 36.5 mm x 30.5 mm and height of 1.59 mm is used. The patch dimension is 11.86 mm x 9.06 mm for Rogers RT/duroid 5880.All values for different substrate materials are calculated using equation that are described in above discussion. The feed location is kept variable.

SIMULATION RESULTS

The Simulation results of proposed antennas are performed by HFSS™. HFSS™ stands for High Frequency Structure Simulator. HFSS™ is a high-performance full- wave electromagnetic (EM) field simulator for arbitrary 3D volumetric passive device modeling that takes advantage of the familiar Microsoft Windows Graphical User Interface (GUI).
It integrates simulation, visualization, solid modeling, and automation in an easy-to-learn environment where solutions to your 3D EM   problems   are quickly and accurately obtained. Ansoft HFSS™ employs the Finite Element Method (FEM), adaptive meshing, and brilliant graphics to give you unparalleled performance and insight to all of your 3D EM problems. Ansoft HFSS™ can be used to calculate parameters such as S-Parameters, Resonant Frequency, and Fields [11].

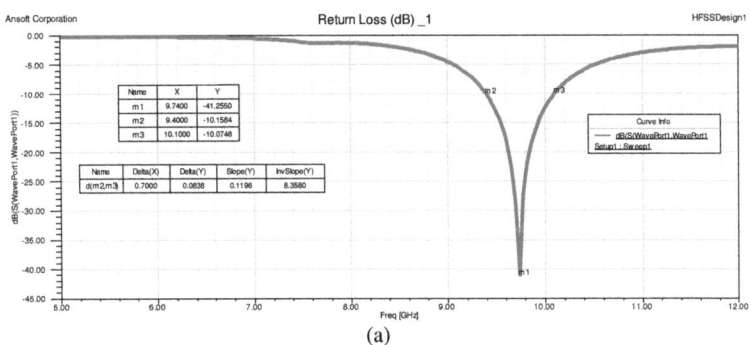

(a)

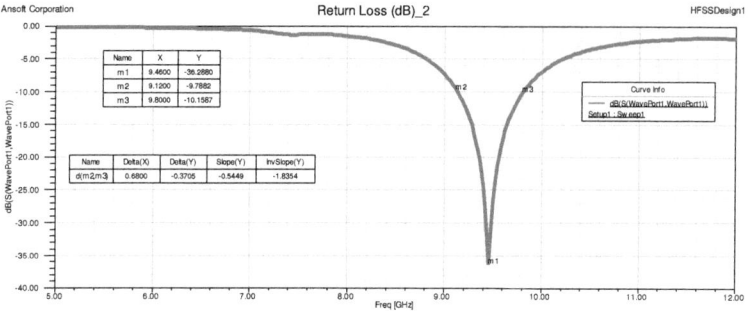

(b)

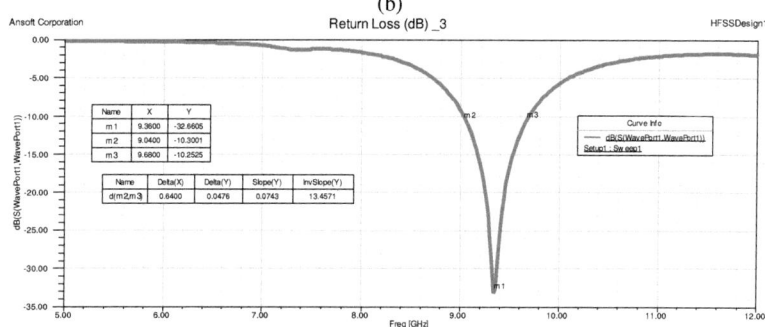

(c)

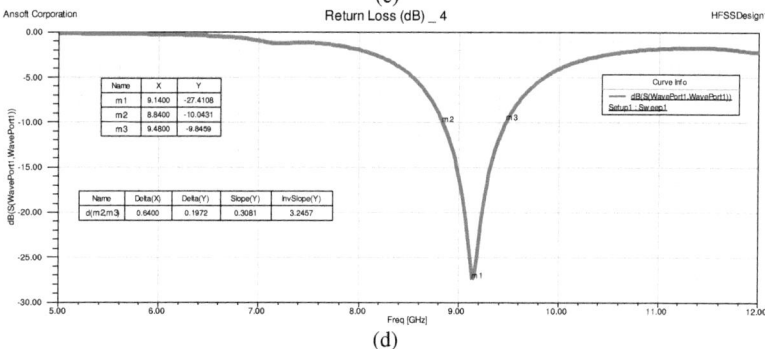

(d)

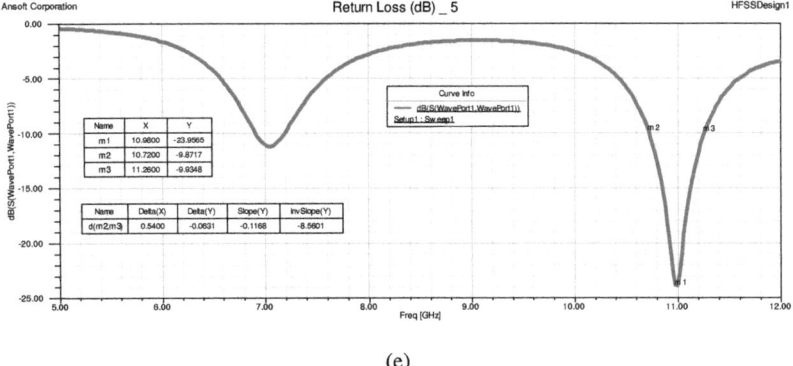

(e)

Figure 2: Return loss vs. Frequency plots of RMSA with (a) Rogers RT/duroid 5880 (b) Rogers RT/duroid 5870 (c) Neltec NX9240 (d) Arlon DiClad 522 (e) FR4_epoxy

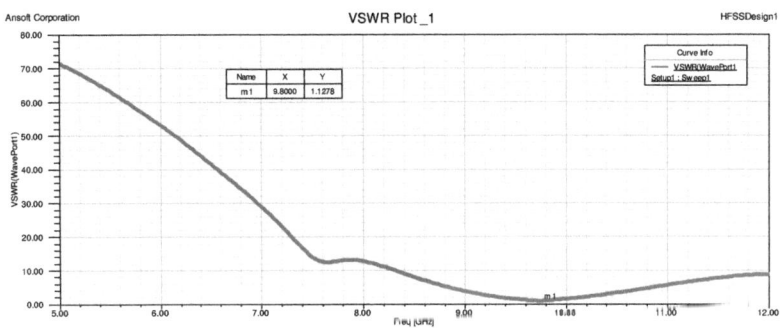

(a)

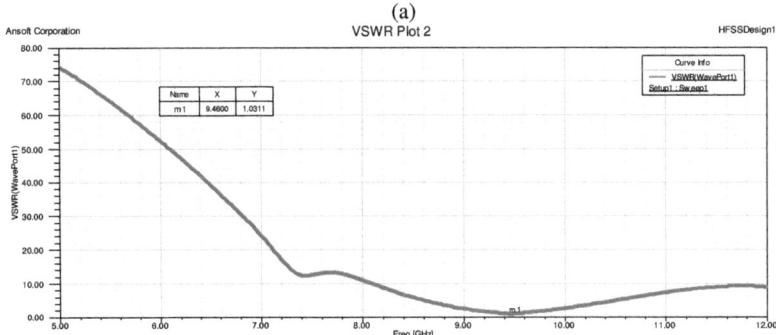

(b)

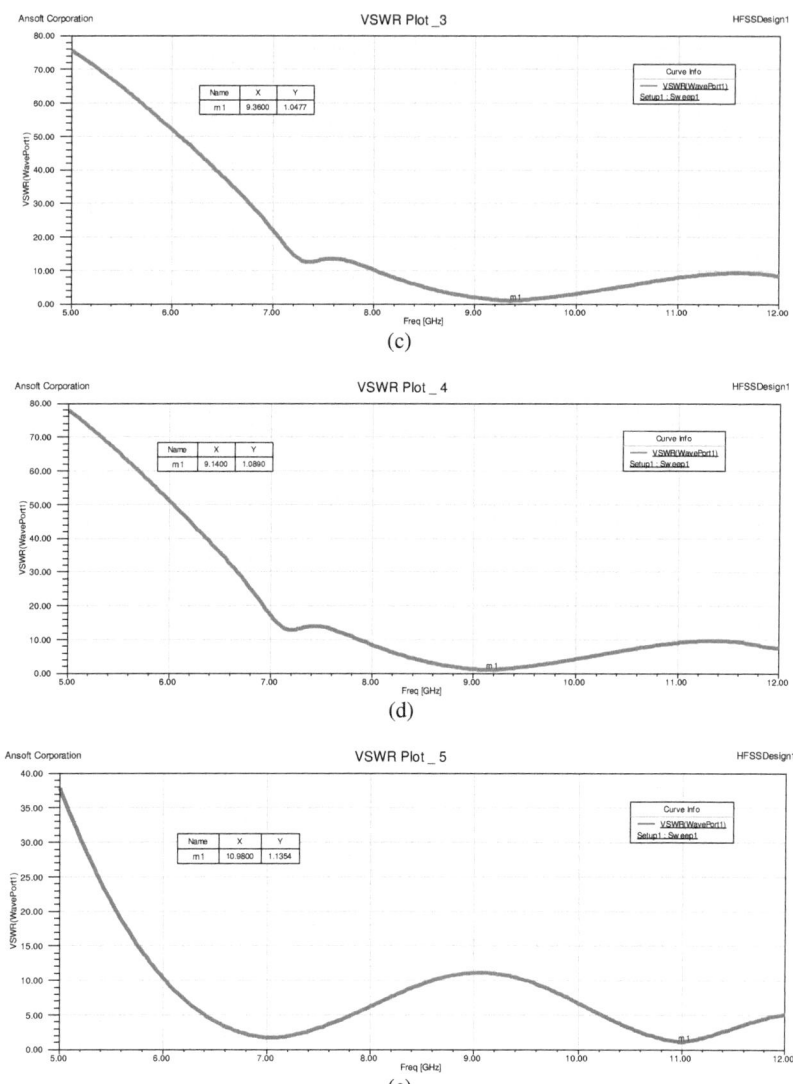

Figure 3: Voltage standing wave ratio (VSWR) plots of RMSA with (a) Rogers RT/duroid 5880 (b) Rogers RT/duroid 5870 (c) Neltec NX9240 (d) Arlon DiClad 522 (e) FR4_epoxy

Figure 2 shows return loss vs. frequency plots of RMSA using different substrates with different dielectric constant. Figure 3 shows VSWR plots of RMSA with different substrates. VSWR < 2 is obtained at the resonant frequency for all the antennas. Figure 4 shows radiation patterns at phi=0 deg (E-plane) and at phi=90 deg (H-plane) for all antennas. Figure 5 shows 2D gaintotal (in dB) plots of all antennas. Figure 6 shows 3D polar plots for total gain (in dB) and figure 7 shows smith charts for input impedance for all antenna using different substrate materials.

Radiation Pattern 1

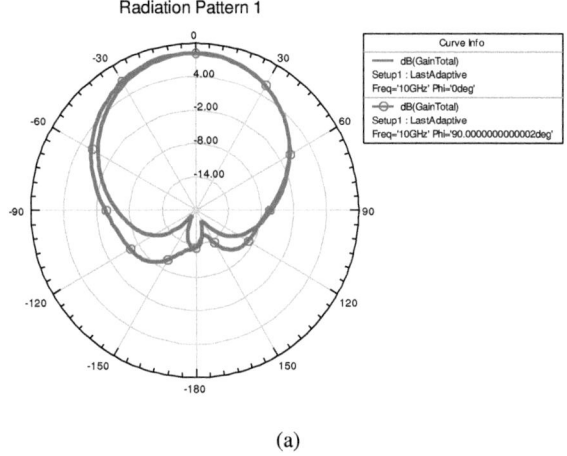

(a)

Radiation Pattern 2

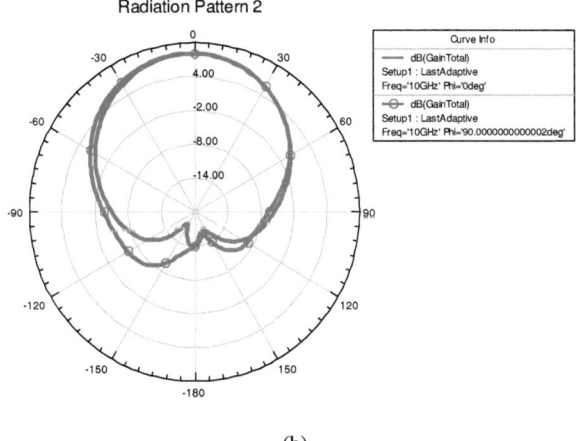

(b)

Radiation Pattern 3

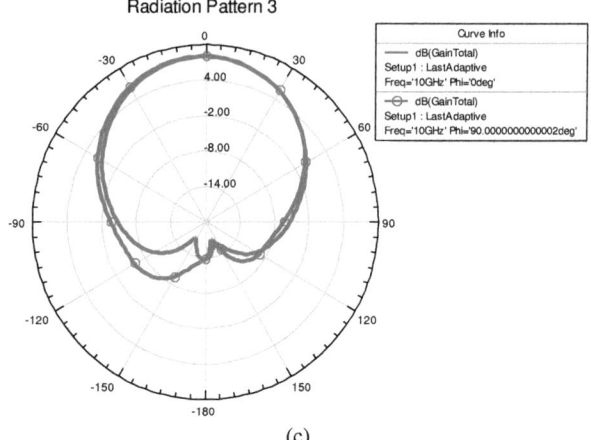

(c)

Radiation Pattern 4

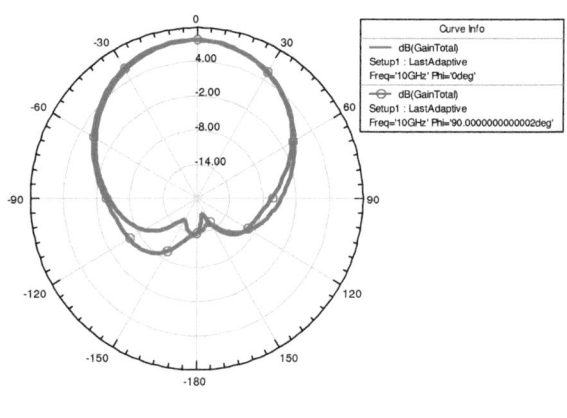

(d)

Radiation Pattern 5

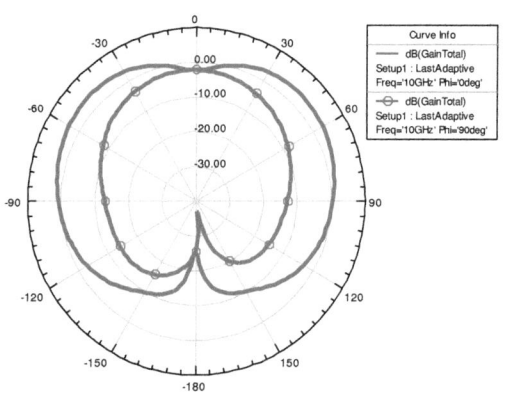

Figure 4: Radiation pattern plots at phi=0 deg (E-plane) and phi= 90 deg (H-plane) of RMSA with (a) Rogers RT/duroid 5880 (b) Rogers RT/duroid 5870 (c) Neltec NX9240 (d) Arlon DiClad 522 (e) FR4_epoxy

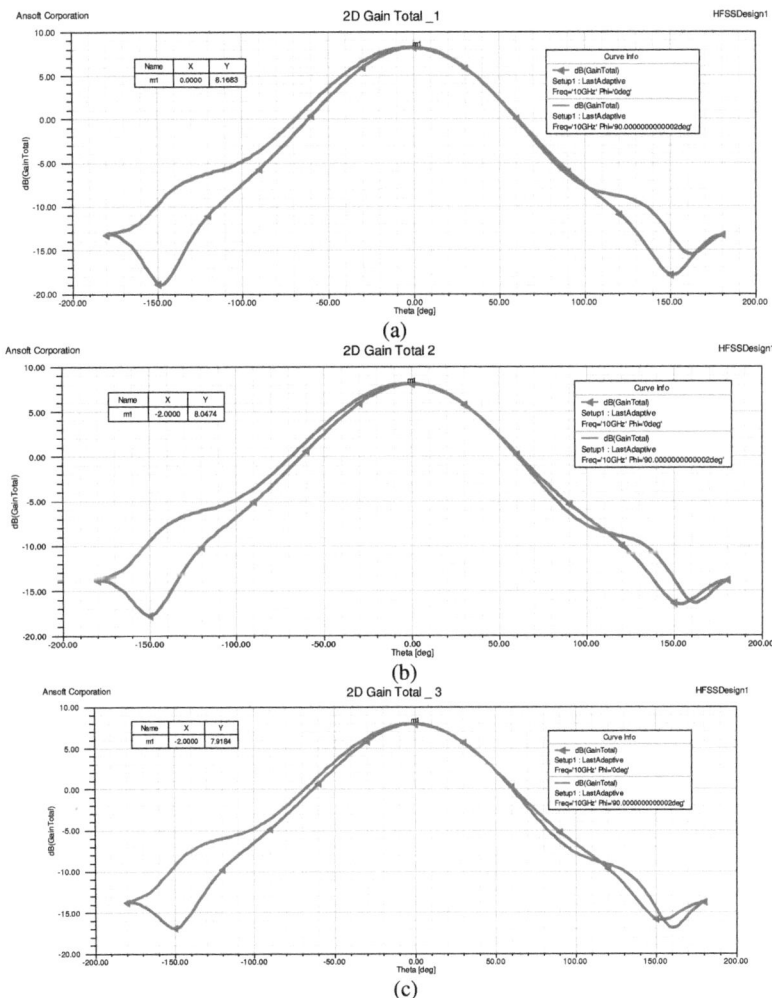

(a)

(b)

(c)

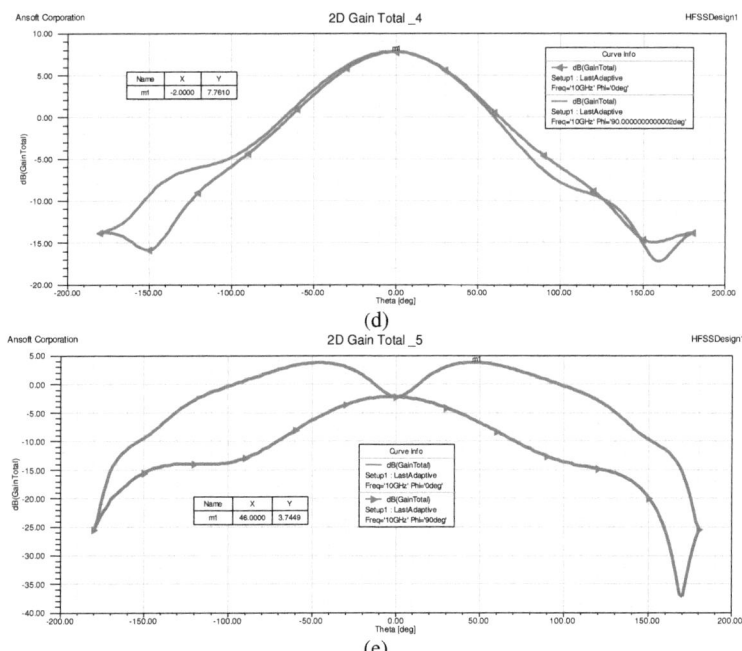

(d)

(e)

Figure 5: 2D gaintotal plots of RMSA with (a) Rogers RT/duroid 5880 (b) Rogers RT/duroid 5870 (c) Neltec NX9240 (d) Arlon DiClad 522 (e) FR4_epoxy

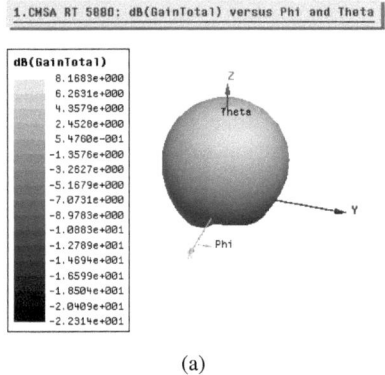

(a)

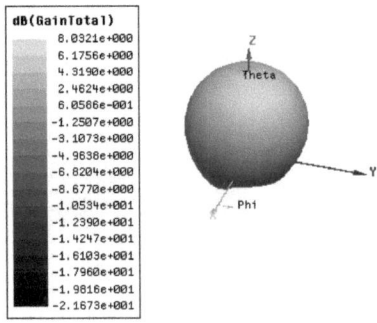

(b)

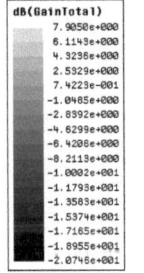
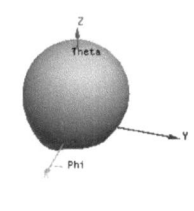

(c)

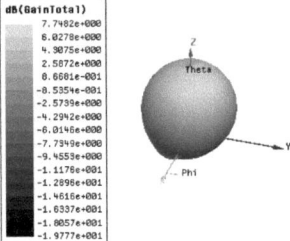

(d)

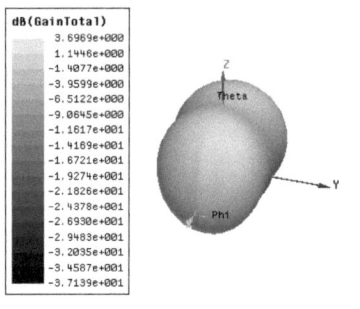

(e)

Figure 6: 3D Polar plots for total gain of RMSA with (a) Rogers RT/duroid 5880 (b) Rogers RT/duroid 5870 (c) Neltec NX9240 (d) Arlon DiClad 522 (e) FR4_epoxy

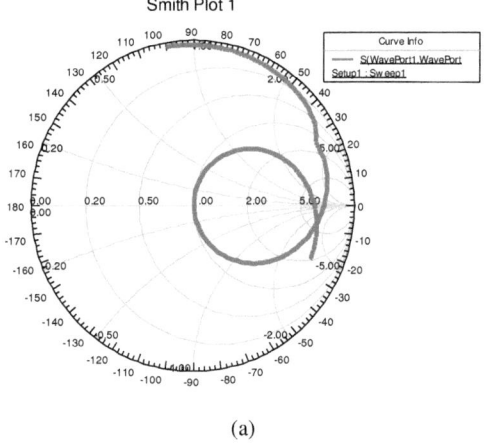

(a)

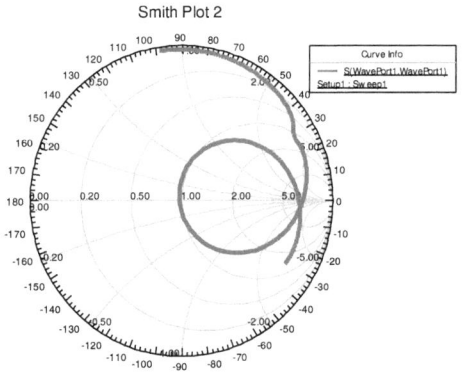

(b)

Smith Plot 3

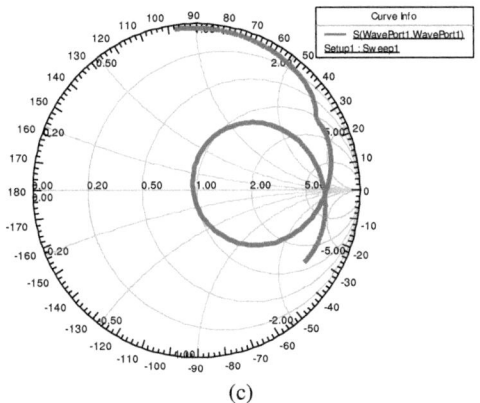

(c)

Smith Plot 4

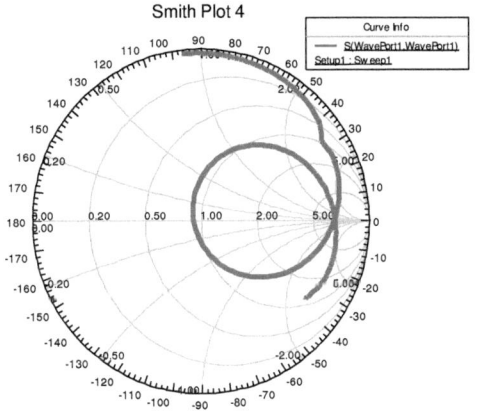

(d)

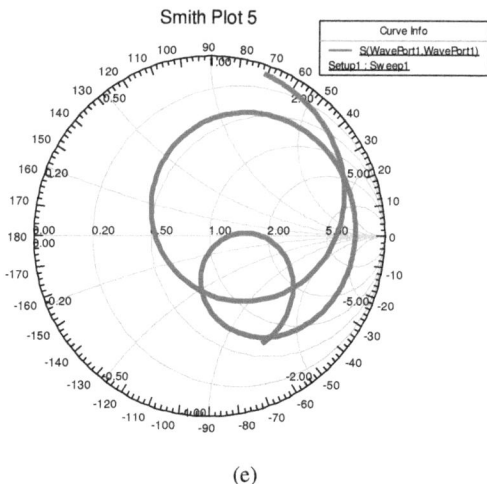

Smith Plot 5

(e)

Figure 7: Smith Chart for Input Impedance of RMSA with (a) Rogers RT/duroid 5880 (b) Rogers RT/duroid 5870 (c) Neltec NX9240 (d) Arlon DiClad 522 (e) FR4_epoxy

As shown in figure 1 and 2 RMSA with substrate Rogers RT/duroid 5880 gives return loss of -41.25 dB with 700 MHz bandwidth and VSWR is 1.01, Rogers RT/duroid 5870 gives return loss of -36.28 dB with 680 MHz bandwidth and VSWR is 1.03, Neltec NX9240 gives return loss of -32.66 dB with 640 MHz bandwidth and VSWR is 1.04,Arlon DiClad 522 gives return loss of -27.41 dB with 640 MHz bandwidth and VSWR is 1.08, and FR4_epoxy gives return loss of -23.95 dB with 540 MHz bandwidth and VSWR is 1.13. The simulation result of all designed and simulated antennas is given in Table 2.Table 3 gives input impedance parameters. Antenna parameters at 10 GHz are tabulated in Table 4.

CONCLUSION

A rectangular microstrip patch antennas (RMSAs) using different substrate materials is designed and its performance at X-band frequency is analyzed. Substrate with lower dielectric constant (εr) value can increase the fringing field in patch and in turn the radiated power and antenna efficiency. Dielectric loss increases with high loss tangent (tan δ) and it reduces antenna efficiency. In this research article different substrate materials including Rogers RT/duroid 5880, Rogers RT/duroid 5870, Neltec NX9240, Arlon DiClad 522, and FR4_epoxy are analyzed with rectangular microstrip patch antenna configuration. Rogers RT/duroid 5880 substrate gives good return loss and VSWR with better impedance bandwidth. A maximum gain of 8.16 dB with better return loss of -41.25 dB is obtained for the Rogers RT/duroid 5880 substrate compared to all other antennas.

Substrate Material	Resonant Frequency f_r (GHz)	Dielectric constant (ε_r)	Dielectric Loss tangent (tan δ)
Rogers RT/duroid 5880	10	2.2	0.0009
Rogers RT/duroid 5870	10	2.3	0.0012
Neltec NX9240	10	2.4	0.0016

Arlon DiClad 522	10	2.5	0.0010
FR4_epoxy	10	4.4	0.02

Table 1: Substrate materials with dielectric constant

Substrate Material	Resonant Frequency f_r (GHz)	Height of Substrate (mm)	Return Loss (dB)	VSWR	Return Loss Bandwidth (GHz)	Gain (dB)	Input Impedance (ohm)
Rogers RT/duroid 5880	10	1.59	-41.25	1.01	0.700	8.16	75.23
Rogers RT/duroid 5870	10	1.59	-36.28	1.03	0.680	8.04	75.27
Neltec NX9240	10	1.59	-32.66	1.04	0.640	7.91	75.23
Arlon DiClad 522	10	1.59	-27.41	1.08	0.640	7.76	75.23
FR4_epoxy	10	1.59	-23.95	1.13	0.540	3.74	75.25

Table 2: Results of RMSA with different substrate materials

Parameter	Rogers RT/duroid 5880	Rogers RT/duroid 5870	Neltec NX9240	Arlon DiClad 522	FR4_epoxy
rms value	0.7734	0.7739	0.7749	0.7755	0.6820
Gain margin	35.90	35.87	32.87	27.10	10.23
Phase margin	279.91	278.45	277.65	276.54	250.99
Gain cross over	5.00	5.00	5.00	5.00	5.00
Phase cross over	9.50	9.45	9.35	9.15	7.17
Upper cut off	8.73	8.50	8.41	8.24	6.33
Band width	8.73	8.50	8.41	8.24	6.33

Table 3: Input Impedance parameter at 10 GHz

Antenna	Rogers RT/duroid 5880	Rogers RT/duroid 5870	Neltec NX9240	Arlon DiClad 522	FR4_epoxy
Quantity	Value (Unit)	Value (Unit)	Value (Unit)	Value (Unit)	Value (Unit)
Max U	0.49819 (W/sr)	0.40833 (W/sr)	0.3636 (W/sr)	0.29392 (W/sr)	0.085251 (W/sr)
Peak Directivity	6.6019	6.3868	6.2495	6.0313	3.5488
Peak gain	6.6244	6.4211	6.2382	6.0198	2.381
Peak realized gain	6.2606	5.1313	4.5702	3.6936	1.0731
Radiated power	0.94727 (W)	0.80342 (W)	0.73129 (W)	0.6124 (W)	0.30189 (W)
Accepted power	0.94508 (W)	0.79914 (W)	0.73261 (W)	0.61357 (W)	0.44995 (W)
Incident power	1 (W)	1 (W)	1 (W)	1 (W)	1 (W)
Radiation efficiency	1.00	1.00	0.99	0.99	0.67093
Front to back ratio	161.53	173.57	163.7	162.74	8.97

Table 4: Antenna Parameters at 10 GHz

REFERENCES

[1] Kin-Lu Wong, "Compact and Broadband Microstrip Antennas", A Wiley-Interscience Publication, John Wiley & Sons, Inc., 2002

[2] Fang, D. G. "Antenna theory and microstrip antennas", CRC Press Taylor & Francis Group, 2010.

[3] Ramesh Garg, Prakash Bhartia,Inder J. Bahl and Apisak Ittipiboon ,"Microstrip Antenna Design   Handbook", Artech House,Inc.,Boston,London, 2001.

[4] James R. James, Peter S. Hall, "Handbook of Microstrip Antennas", Peter Peregrinus Ltd., London, UK, IEE Electromagnetic Waves Series 28, 1989.

[5] Keith R. Carver and James Mink, "Microstrip Antenna Technology", IEEE Transactions on Antennas and Propagation Vol. 29, No. 1, pp. 2-22, 1981.

[6] T. E. Nowicki, "Microwave substrates, present and future," in Proc. Workshop Printed Circuit Antenna Tech., New Mexico State Univ., Las Cruces, Oct. 1979, pp. 26/1-22.

[7] Constantine A. Balanis, "Antenna Theory, Analysis and Design", 2nd ed., John Wiley & Sons, Inc., New York, 1997.

[8] D. H. Schaubert, D. M. Pozar, and A. Adrian, "Effect of Microstrip Antenna SubstrateThickness and Permittivity: Comparison of Theories and Experiment," IEEE Trans. Antennas Propagatation., Vol. AP-37, No. 6, pp. 677–682, June 1989

[9] E. O. Hammerstad, "Equations for Microstrip Circuit Design," Proc. Fifth European Microwave Conf., pp. 268–272, September 1975.

[10] C. A. Balanis, "Advanced Engineering Electromagnetics", JohnWiley & Sons, New York, 1989.

[11] Ansoft High Frequency Structure Simulator v11 and V13 User's Guide, REV1.0, 2005 Ansoft Corporation.